The Thomas D. Clark Lectures

1990

THE REDISCOVERY OF NORTH AMERICA

Barry Lopez

THE UNIVERSITY PRESS OF KENTUCKY

Publication of this book was assisted by a grant
from the Gaines Center for the Humanities, which
initiated and supports the Thomas D. Clark
Lectureship Series.

The author would like to express his appreciation to
Robley Wilson for his help with the manuscript.

Published by The University Press of Kentucky

Scholarly publisher for the Commonwealth,
serving Bellarmine College, Berea College, Centre
College of Kentucky, Eastern Kentucky University,
The Filson Club, Georgetown College, Kentucky
Historical Society, Kentucky State University,
Morehead State University, Murray State University,
Northern Kentucky University, Transylvania University,
University of Kentucky, University of Louisville,
and Western Kentucky University.

Editorial and Sales Offices: Lexington, Kentucky 40508-4008

Second Printing

Library of Congress Cataloging-in-Publication Data

Lopez, Barry Holstun, 1945–
    The rediscovery of North America / Barry Lopez.
        p.   cm. — (The Thomas D. Clark lectures : 1990)
    ISBN 0-8131-1742-9
    1. North America—Description and travel—1981– 2. Man—Influence
on nature—North America. 3. America—Discovery and exploration
—Spanish—Influence. I. Title. II. Series.
E27.5.L67   1990
970.01'6—dc20   90-24487

Dedicated to the memory of
Rachel Carson
(1907-1964)

THE REDISCOVERY OF
NORTH AMERICA

A FEW HOURS after midnight on the morning of October twelfth in the Julian calendar of the West—or October twenty-second, according to the modern Gregorian calendar—Juan Rodriguez Bermeo, a lookout aboard the caravel *Pinta*, spotted the coast of either San Salvador Island or Samana Cay in the Bahamas and shouted his exclamation into the darkness. It was the eighteenth year of the reign of Ferdinand and Isabella of Castile, and these mariners were their emissaries.

Cristoforo Colombo—or Christopher Dove as it would be in English—commander of the fleet of three ships, gave orders to take in sail, and to lay close-hauled five miles off shore awaiting the rise of the sun. The seas were rolling.

Strong winds tore at the crests of the waves. A gibbous moon was setting in a clear sky.

As they awaited dawn, Columbus let it be known that he had earlier seen a light on the island, a few hours before midnight. The ships were making about ten knots when Bermeo cried out. By his claim the commander would had to have seen the light at a distance of more than thirty miles over the curve of the Earth. Columbus thereby took for himself the lifetime pension promised the first man to sight land.

Of Señor Bermeo history has little more to say. It was rumored that he converted to Islam and died fighting alongside the Moors, who had that year of

1492 lost their final stronghold in Spain, in the same year Jews were evicted from the country by royal edict.

We do not know what Columbus and his men envisioned when they came ashore on Samana Cay or San Salvador, the island the local Arawak people called Guanahaní, in a chain the Spanish were to call the Lucayas. But we know that in those first few hours a process began we now call an incursion. In the name of distant and abstract powers, the Spanish began an appropriation of the place, a seizure of its people, its elements, whatever could be carried off.

What followed for decades upon this discovery were the acts of criminals—murder, rape, theft, kidnapping, vandalism, child molestation, acts of cruelty, torture, and humiliation. Bartolomé de las Casas, who arrived in Hispaniola in 1502 and later became a priest, was an eyewitness to what he called "the obdurate and dreadful temper" of the Spanish, which "attended [their] unlimited and close-fisted avarice," their vicious search for wealth. One day, in front of Las Casas, the Spanish dismembered, beheaded or raped three thousand people. "Such inhumanities and barbarisms were committed in my sight," he says, "as no age can parallel. . . ." The Spanish cut off the legs of children who ran from them. They poured people full of boiling soap. They made bets as to who, with one sweep of his sword, could cut a person in half. They loosed dogs that "devoured an Indian like a hog, at first sight, in less than a moment." They used nursing infants for dog food.

It was "a continuous recreational slaughter," practiced by men who felt slights to their personages, imagined insults to their religion, or felt thwarted in their search for gold or sexual congress.

These words of Las Casas's—who said "I resolve silently to pass over, lest I should terrify the reader with the horror," a more graphic recounting of these incidents—were written at Valencia in 1542 at the request of historians, "to display to the world the enormities, etc., [that] the Spaniards committed in America to their eternal ignominy." Las Casas writes in the opening pages of this treatise, "I earnestly beg and desire all men to be persuaded that this summary was not published upon any private design, sinister ends or affection in favor of or prejudice of any particular nation; but for the public emolument and advantage of all true Christians and moral men throughout the world."

I SINGLE OUT these episodes of depravity not so much to indict the Spanish as to make two points. First, this incursion, this harmful road into the "New World," quickly became a ruthless, angry search for wealth. It set a tone in the Americas. The quest for personal possessions was to be, from the outset, a series of raids, irresponsible and criminal, a spree, in which an end to it—the slaves, the timber, the pearls, the fur, the precious ores, and, later, arable land, coal, oil, and iron ore—was never visible, in which an end to it had no meaning.

The assumption of an imperial right conferred by God, sanctioned by the state, and enforced by a militia; the assumption of unquestioned superiority over a resident people, based not on morality but on race and cultural comparison—or, let me say it plainly, on ignorance, on a fundamental illiteracy—the assumption that one is *due* wealth in North America, reverberates in the journals of people on the Oregon Trail, in the public speeches of nineteenth-century industrialists, and in twentieth-century politics. You can hear it today in the rhetoric of timber barons in my home state of Oregon, standing before the last of the old-growth forest, irritated that anyone is saying *"enough . . . , it is enough."*

What Columbus began, then, what Pizarro and Cortés and Coronado perpetuated, is not isolated in the past. We see a continuance in the present of this brutal, avaricious behavior, a profound abuse of the place during the course of centuries of demand for material wealth. We need only look for verification at the acid-burned forests of New Hampshire, at the cauterized soils of Iowa, or at the collapse of the San Joaquin Valley into caverns emptied of their fossil waters.

The second point I wish to make is that this violent corruption needn't define us. Looking back on the Spanish incursion, we can take the measure of the horror and assert that we will not be bound by it. We can say, yes, this happened, and we are ashamed. We repudiate the greed. We recognize and condemn the evil. And we see how the harm has been perpetuated. But, five hundred years later, we intend to mean something else in the world.

IMAGINE GUANAHANÍ, where Columbus put ashore that windy Friday morning. It is the most seaward of the Bahamas. If we mean, specifically, San Salvador, it is today a quiet, sparsely inhabited island with a small village, Cockburn Town, and up the road a single hotel visited mostly by divers, who come here to see an impressive array of tropical fish, of underwater plants and creatures, and, just offshore, massive walls where the bottom drops off into a bottomless blue.

The island itself has changed dramatically since Columbus arrived. Its surface, in essence, was scraped bare and then, in the centuries that followed, it was repopulated by refugees—people of complex origin, a mixture of local and foreign plants and animals. The same is true across much of the Caribbean. We cannot today see what Columbus saw; that fabric of animals and plants and people was utterly torn apart.

It is hard to conceive how amazed the Spanish must have been in those first few years by the sight of crocodiles, hummingbirds, and the roseate spoonbill; the anaconda and the jaguar; rhinoceros beetles and howler monkeys. But these, ultimately, were only beautiful distractions. The conquistadors would throw away this splendor in a moment, and did, for the silver of Potosí. Had the Maryland darter, the dusky seaside sparrow, the Palos Verdes blue butterfly, or the plains grizzly been able to read, they would have seen their own fate sealed here.

The Spanish sought a narrowly defined wealth. Las Casas writes, "Now the ultimate end and scope that incited the Spaniards to endeavor the extirpation and desolation of this people was gold only; that thereby growing opulent in a short time they might arrive at once at such degrees and dignities as were in no ways consistent with their persons."

We lost in this manner whole communities of people, plants, and animals, because a handful of men wanted gold and silver, title to land, the privileges of aristocracy, slaves, stables of little boys. We lost languages, epistemologies, books, ceremonies, systems of logic and metaphysics—a long, hideous carnage.

It is perilous, of course, to suggest

that we ourselves would have behaved differently. (My generation turns back but a few pages to scenes in the villages of Vietnam, where our goals were simply political.) I also mean to take the long view here. But it is not good to forget, not to face squarely, what happened, the way the world forgot the extermination of Armenians in Turkey twenty-five years before Buchenwald. And if we say, yes, all right, this was us, and the pattern continues, then how are we to understand it? How can we clarify for ourselves what went wrong? How can we claim not that we are different but that we wish, seeing what has come in the wake of our acts, to set off now in a different direction?

A distinction that is crucial and instructive in the face of this dilemma is suggested by Tzvetan Todorov, the French writer and critic, in *The Conquest of America*. He says that what we see in the New World under the Spanish is an imposition of will. It is an incursion with no proposals. The Spanish *impose*, they do not *propose*. I think it is possible to view the entire colonial enterprise, beginning in 1492, in these terms. Instead of an encounter with "the other" in which we proposed certain ideas, proposals based on assumptions of equality, respectfully tendered, our encounters were distinguished by a stern, relentless imposition of ideas—religious, economic, and social ideas we deemed superior if not unimpeachable.

Our trouble with the New World—a

world that was intended to refuel an Old World which had in some sense grown effete—has been that from the beginning we have imposed, not proposed. We never said to the people or the animals or the plants or the rivers or the mountains: What do you think of this? We said what *we* thought, and bent to our will whatever resisted. I do not suggest lightly, or as a kind of romance, that we might have addressed the animals, the trees, the land itself. The idea of this kind of courtesy is more ancient than "primitive." And the wisdom of it, the ineffable and subtle intertwining of living organisms on the Earth, is confirmed today by molecular biology and atmospheric chemistry. To acknowledge the interdependence is simply a good and wise habit of mind.

When we arrived in the New World, we came to talk, not to listen. Now that we have begun to listen to the land, to take into account in our planning the biological and chemical responses of a particular landscape, what we are hearing is the voice that answered Juan Rodriguez Bermeo but which was never heeded. In the beginning it was an antiphony we wanted no part of. We're anxious now to know what the land has to say to us, how it responds to our use of it. And we are curious, too, about indigenous systems of natural philosophy, how our own Western proposals might be answered by some bit of this local wisdom, an insight into how to conduct our life here so that it might be richer. And so that what is left of what we have subjugated might determine its own life.

WHAT WEALTH did Bermeo cry out in anticipation of? Let us be kind, as we hope historians will be kind to us, and say that for him and for many others on those three small ships perhaps the hummingbird would have been enough. The hummingbird, fresh food after five weeks at sea, and the astonishing lives of the Arawak. But the record tells us that in the end there was very little imagination here—it was gold, silver, pearls, slaves, and sexual intercourse. It was venal greed, a failure of imagination, the reduction of desire to its most banal elements. True wealth—sanctity, companionship, wisdom, joy, serenity—these things were not to be had without an offer of heart and soul and time. The Spaniards had no time, and we find it easy to say on the evidence that they were heartless and immoral. The only wealth they *could* imagine was what they took.

The Spanish wanted no communion with America, the place or its people. Residence, except residence construed as land ownership, was not of interest to them. America was not to be a home or what a home implied—the responsibilities and obligations of adult life. They had left that behind in Europe, had traded it away for lawlessness. If we say that the elements of true wealth come with the maintenance of a home, as I think is possible, then we have to say that the Spanish and their descendants were not to find true wealth in America until they discovered the America they had missed.

The true wealth that America offered, wealth that could turn exploitation into residency, greed into harmony, was to come from one thing—the cultivation and achievement of local knowledge. It was in the pursuit of local knowledge alone that one could comprehend the notion of a home and its attendant responsibilities. So the first questions at Guanahaní might better have been: Who are these people? What is this land?

WE RARELY THINK any more of this geography, of how the land was peopled. Start in the Bahamas, with an Arawak people, the Lucayo. Then turn up north and west, to the mainland, to the country of the Calusa. And on north, to the separate landscapes of the Apalachee, the Creek, the Chicasaw. Then east and north, the Cherokee, Delaware, Susquehanna; the Onondaga, Mahican, and Abenaki. Back west: Huron, Ottawa, Chippewa, Plains Ojibwa, Assiniboin, Crow, Shoshone. Into the Northwest: Salish, Nez Perce, Coeur d'Alene, Yakima, Snuqualmi, Chehalis. Down the coast: Tillamook, Siuslaw, Coos, Yurok, Pomo, Maidu, Chumash. Back to the east: Paiute, Havasupai, Papago, Pima, Chiricahua

Apache, Mescalero Apache; Tiwa, Tano, Tewa, Keres, Santo Domingo, Pojoaque, Jicarilla Apache, Kiowa, Arapaho, Pawnee, Kansa, Osage, Shawnee. And back south: Catawba, Hitchiti, Timucua.

This is to leave out all the tribes of the North—Dogrib, Hare, Kutchin; the Northwest Coast—Tlingit and Kwakiutl; of Mexico—Tarahumara, Aztec, Maya; and to the south of them Mosquito and Cuna. And all of South America, from the Carib and Timote to the Yahgan and Ona.

More than a thousand distinct cultures, a thousand mutually unintelligible languages, a thousand ways of knowing. How can one compare intimacy with the facets of this knowledge to the possession of gold? How could we have squandered such wisdom in that search?

Imagine the physical place. It is not often, either, that we call this to mind. Start in the same region. The Florida Keys, that white light. Okefenokee Swamp. The Piedmont Plateau. James River. The pine barrens of New Jersey. Hardwood forests of the White Mountains. Mt. Katahdin. The Finger Lakes. Allegheny Mountains. Cumberland Plateau. Kentucky River. Indiana dunes. Wisconsin dells. Minnesota lakes. Laurentian Divide. Shortgrass prairie. Missouri breaks. Judith Basin. Middle Fork of the Salmon. Mt. Rainier. Olympic Peninsula. Columbia bar. California redwoods. Tuolumne meadows. Black Rock Desert. The

Wasatch Range. Medicine Bow Mountains. South Fork of the Platte. Front Range. Headwaters of the Canadian. Sangre de Cristo. Mogollon Rim. Sonoran Desert. Edwards Plateau. Brazos River. Big Thicket. Lake Pontchartrain. Chattahoochee River. Everglades.

And I leave out, again, all that to the north: Thelon River, Great Bear Lake, the Brooks Range, Glacier Bay. And to the south: the Chihuahuan Desert, Yucatán Peninsula, Mosquito Coast, the Darien Isthmus. And South America: Lake Titicaca, Mato Grosso, Atacama Desert. Patagonia. Tierra del Fuego.

It would take a lifetime to list, even in these few places, the trees and flowers, the butterflies and fish, the small mammals,

the kinds of deer and cats, the migratory
and resident birds; and to say the most
rudimentary things about their relation-
ships, how they know and reflect each
other.

This, along with the people we ig-
nored, was a wealth that didn't register
until much of it was gone, or until, like the
people, it was a tattered, diluted remnant,
sequestered on a reservation.

This brief litany of names, though all
are imposed, should awaken a sense of the
breadth of this landscape, of how by turns
it is strange and comprehensible, familiar
and unfathomable. It is still in some real
sense the New World. These are still
places with which we might reciprocate.

CAMUS SAID THAT certain cities—he had in mind Oran on the Algerian coast—exorcised the landscape. We have a way of life that ostracises the land. Cadwallader Colden, John Bartram, Peter Collinson, Mark Catesby, Thomas Say, John Kirk Townsend, Thomas Nuttall, John James Audubon—all were able at least to describe what they found. But this extensive knowledge was ultimately regarded as only a kind of entertainment. Decorative information. A series of puzzles for science to elucidate. It was never taken to be what it in fact is—a description of home.

How, then, do we come to know the land, to discover what more may be there than merchantable timber, grazeable prairies, recoverable ores, damable water, netable fish?

It is by looking upon the land not as its possessor but as a companion. To achieve this, one must I think cultivate intimacy, as one would with a human being. And that would mean being *in* a place, taking up residence in a place. Let us choose, for an example, a residence in the basin of the Kentucky River, a tributary of the Ohio. Because we ourselves are recent arrivals, we would have to read local history, would have to find a memory of the place through the journals and records of those of us who first came across the Alleghenies and over the Cumberland Plateau—Thomas Walker through the Cumberland Gap for the

Loyal Land Company, and residents of the proposed colonies of Transylvania and Vandalia—would have to read these observations against each other. And then read in the anthropological and archeological literature about those we moved out of our way—the Shawnee and Miami; and about those *they* followed here, the Hopewell; and those *they* followed, the Adena—as far back as we could go. And then we would come forward, looking in the archives of towns and countries for records of observation that spanned enough years so that the human span of three score and ten was not the span we measured the country by.

We would have to memorize and remember the land, walk it, eat from its soils

and from the animals that ate its plants.
We would have to know its winds, inhale
its airs, observe the sequence of its flowers
in the spring and the range of its birds.

To enquire after this knowledge is to
make our proposals, to answer the antiphony. To be intimate with the land like
this is to enclose it in the same moral
universe we occupy, to include it in the
meaning of the word community.

IT HAS BEEN my privilege to travel, to see a lot of country, and in those travels I have learned of several ways to become intimate with the land, ways I try to practice. I remember a Nunamiut man at Anaktuvuk Pass in the Brooks Range in Alaska named Justus Mekiana. I was there working on a book and I asked him what he did when *he* went into a foreign landscape. He said, "I listen."

And a man named Levine Williams, a Koyukon Athapaskan, who spoke sternly to a friend, after he had made an innocent remark about how intelligent people were, saying to him, "Every animal knows way more than you do."

And another man, an Inuk, watching a group of polar bear biologists on Baffin

Island comparing notes on the migration paths of polar bears, in an effort to predict where they might go. "Quajijau-jungangitut," he said softly, "it can't be learned."

I remember a Kamba man in Kenya, Kamoya Kimeu, a companion in the stone desert west of Lake Turkana—and a dozen other men—telling me, you know how to see, learn how to *mark* the country. And he and others teaching me to sit down in one place for two or three hours and look.

When we enter the landscape to learn something, we are obligated, I think, to pay attention rather than constantly to pose questions. To approach the land as we would a person, by opening an intelligent conversation. And to stay in one place, to make of that one, long observation a fully dilated experience. We will always be rewarded if we give the land credit for more than we imagine, and if we imagine it as being more complex even than language.

In these ways we begin, I think, to find a home, to sense how to fit a place.

IN SPANISH, *la querencia* refers to a place on the ground where one feels secure, a place from which one's strength of character is drawn. It comes from the verb *querer*, to desire, but this verb also carries the sense of accepting a challenge, as in a game.

In Spain, *querencia* is most often used to describe the spot in a bullring where a wounded bull goes to gather himself, the place he returns to after his painful encounters with the picadors and the banderilleros. It is unfortunate that the word is compromised in this way, for the idea itself is quite beautiful—a place in which we know exactly who we are. The place from which we speak our deepest beliefs. *Querencia* conveys more than "hearth."

And it carries this sense of being challenged—in the case of a bullfight, by something lethal, which one may want no part of.

I would like to take this word *querencia* beyond its ordinary meaning and suggest that it applies to our challenge in the modern world, that our search for a *querencia* is both a response to threat and a desire to find out who we are. And the discovery of a *querencia*, I believe, hinges on the perfection of a sense of place.

A sense of place must include, at the very least, knowledge of what is inviolate about the relationship between a people and the place they occupy, and certainly, too, how the destruction of this relationship, or the failure to attend to it, wounds people. Living in North America and trying to develop a philosophy of place—a recognition of the spiritual and psychological dimensions of geography—inevitably brings us back to our beginnings here, to the Spanish incursion. The Spanish experience was to amass wealth and go home. Those of us who have stayed, who delight in the litanies of this landscape and who can imagine no deeper pleasure than the fullness of our residency here, look with horror on the survival of that imperial framework in North America—the physical destruction of a local landscape to increase the wealth of people who don't live there, or to supply materials to buyers in distant places who will never know the destruction that process leaves behind.

If, in a philosophy of place, we examine our love of the land—I do not mean a romantic love, but the love Edward Wilson calls biophilia, love of what is alive, and the physical context in which it lives, which we call "the hollow" or "the canebrake" or "the woody draw" or "the canyon"—if, in measuring our love, we feel anger, I think we have a further obligation. It is to develop a hard and focused anger at what continues to be done to the land not so that people can survive, but so that a relatively few people can amass wealth.

I'm aware that these words, or words like them, have historically invoked revolution. But I ask myself, where is the man or woman, standing before lifeless porpoises strangled and bloated in a beach-cast driftnet, or standing on farmland ankle deep in soil gone to flour dust, or flying over the Cascade Mountains and seeing the clearcuts stretching for forty miles, the sunbaked earth, the streams running with mud, who does not want to say, "Forgive me, thou bleeding earth, that I am meek and gentle with these butchers"?

If we ask ourselves what has heightened our sense of loss in North America, what has made us feel around in the dark for a place where we might take a stand, we would have to answer that it is the particulars of what is now called the environmental crisis. Acid rain. Soil erosion. Times Beach. Falling populations of wild animals. Clearcutting. Three Mile Island. But what we really face, I think, is something much larger, something that goes back to Guanahaní and what Columbus decided to do, that series of acts—theft, rape, and murder—of which the environmental crisis is symptomatic. What we face is a crisis of culture, a crisis of character. Five hundred years after the *Niña*, the *Pinta*, and the *Santa Maria* sailed into the Bahamas, we are asking ourselves what has been the price of the assumptions those ships carried, particularly about the primacy of material wealth.

One of our deepest frustrations as a culture, I think, must be that we have made so extreme an investment in mining the continent, created such an infrastructure of nearly endless jobs predicated on the removal and distribution of trees, water, minerals, fish, plants, and oil, that we cannot imagine stopping. In the part of the country where I live, thousands of men are now asking themselves what jobs they will have—for they can see the handwriting on the wall—when they are told they cannot cut down the last few trees and that what little replanting they've done—if it actually works—will not produce enough timber soon enough to ensure their jobs.

The frustration of these men, who are my neighbors, is a frustration I am not deeply sympathetic to—their employers have behaved like wastrels, and they have known for years that this was coming. But in another way I am sympathetic, for these men are trying to live out an American nightmare which our system of schools and our voices of government never told them was ill-founded. There is not the raw material in the woods, or beyond, to make all of us rich. And in striving for it, we will only make ourselves, all of us, poor.

When people have railed against environmentalism for the restrictions it has sought to impose, they have charged—I'm thinking of loggers in Oregon, and shrimp fishermen in the Gulf, and oil drillers on the North Slope—that environmentalists are out to destroy the independent spirit of the American entrepreneur. They've meant to invoke an image of self-reliance and personal responsibility. They've meant by their words to convey this: If something is truly wrong here, we'll see it and fix it. We don't need anyone to tell us what to do.

The deep and tragic confusion here is that this pose of responsibility, this harkening to a heritage of ennobled independence, has no historical foundation in America. Outside of single individuals and a few small groups that attended to the responsibilities of living on the land, attended to the reciprocities involved, the history of the use of the American landscape has been lawless exploitation. When an industry asks to police itself, we must have the courage to note that there is no precedent, that the entrenched precedent, from the time of the Spanish, is lawlessness in the quest for wealth, with the extension of enough local generosity to keep from being run out of town, enough respect for institutions to keep from being hauled before the bar, and enough patriotism to be given the benefit of the doubt by society.

We cannot, with Huck Finn and Mark Twain, light out for the territory any more, to a place where we might continue to live without parental restraint. We need to find our home. We need to find a place where we take on the responsibilities of adults to the human community. Having seen what is going on around us, we need to find, each person, his or her *querencia*, and to believe it is not a matador in a bullring we face, a rigged game, but an assailable beast, another in our history like Tamerlane or the Black Death.

What we need is to discover the continent again. We need to see the land with a less acquisitive frame of mind. We need to sojourn in it again, to discover the lineaments of cooperation with it. We need to discover the difference between the kind of independence that is a desire to be responsible to no one but the self—the independence of the adolescent—and the independence that means the assumption of responsibility in society, the independence of people who no longer need to be supervised. We need to be more discerning about the sources of wealth. And we need to find within ourselves, and nurture, a profound courtesy, an unalloyed honesty.

SOME HOLD THAT this task is hopeless, that the desire for power and wealth is too strong. Without denying in any way the dark flaws in human nature, I wish politely to disagree. I would like to put forth what may pass for sources of hope but which are in fact only examples that we can follow, situations that we can take advantage of, and people who I think might inspire us.

If we are looking for some better way to farm, we need look no further than the Amish and Mennonite communities of the country for that kind of intelligence. And we should remind ourselves that it is not necessary to *be* a people in order to avail ourselves of their intelligence—that in fact such a tack is unwise.

If we are to find examples on which to model our courage, we need look no further than Bartolomé de las Casas, who wrote 450 years ago what is relevant to us today. And if we are afraid of human angels, we need only remind ourselves that Las Casas was, to some extent, also a man of his day. He paid little attention to the plight of black slaves in the New World.

If we would search for a contemporary hero, fighting still this beast the Spanish loosed on these shores, we need only turn our eyes to El Salvador and the murdered archbishop Oscar Romero.

If we require heroes closer to home, people who in their writing, in their essays and meditations, have given us good prescriptions for behavior, we need as a country look no further than the work of Wendell Berry or Thomas Merton.

If we feel wisdom itself is lost, we need only enter a library. We will find there the records of hundreds of men and women who believed in a world larger than the one defined in each generation by human failing. We will find literature, which teaches us again and again how to imagine.

If we become the prisoners of our own minds, if we *think* ourselves into despair, we can step onto wounded ground with a shovel and begin to plant trees. They will grow. They will hold the soil, provide shelter for birds, warm someone's home after we are gone.

If we lose faith in ourselves, we can in

those moments forget ourselves and dwell
on the future of the larger community, on
the blessing of neighbors. Your neighbors
are those you can see when you look out
your window, but today these are not our
only neighbors, if we mean by that word a
common burden, a common joy in an abstract terrain.

IF I THINK back on that long night when the caravels rolled in heavy seas off the coast of Guanahaní, the waning moon setting, the wind blowing hard beneath a clear sky, I can easily imagine men of conscience lying there awaiting the dawn. They could not have known—for they were the first there—what was ahead of them, neither the wonder of it nor how their mettle would be tested.

In a sense we lie there with them. It is our privilege to know what the landscape is actually like—its people, its animals. But we are like them, I think, because we too feel ourselves on the verge of something vague but extraordinary. Something big is in the wind, and we feel it. And we feel, with them, the weight of Columbus's

authority, his compelling political and ec-
clesiastical power. And we sense our own
reluctance, our lack of objection, before
it. His vision, however mad or immoral, is
forceful. Even if we see him as a man
flawed like other men—his megalomania
and delusion, his uncommon longing for
noble titles—we are inclined to see that
he got us across the literally uncharted
ocean, and that that takes a kind of genius.
It puts us, somehow, in his debt. It leaves
room to forgive him, even to believe in his
worthiness. If this search of his for gold
should produce a holocaust, we say to
ourselves, well, then, we might take only
a little, something for our children, a poor
wife waiting at home. And be done with
the man. Who can fight the conviction

that is in Columbus? Who can deny his destiny? Life is short. Let someone in authority take him to task.

We lie in the ships with those men, I think, because we are ambivalent about what to do. We do not know whether to confront this sea of troubles or to stand away, care for our own, and take comfort in the belief that the power to act lies elsewhere.

It is this paralysis in the face of disaster, this fear before the beast, that would cause someone looking from the outside to say that we face a crisis of character. It is not a crisis of policy or of law or of administration. We cannot turn to institutions, to environmental groups, or to government. If we rise in the night, sleepless, to stand at the ship's rail and gaze at the New World under the setting moon, we know we are thousands of miles from home, and that if we mean to make this a true home, we have a monumental adjustment to make, and only our companions on the ship to look to.

We must turn to each other, and sense that this is possible.